by Adrian Harrison

INTRODUCTION TO ROOTS

January 2020

Contents

ROOTS

Definition

$: \sqrt[n]{a} = x \Rightarrow x^n = a$

$\sqrt[n]{a^n} = a$, (n is an odd number)

$\sqrt[n]{a^n} = a$, (n is an even number)

$a < 0 \Rightarrow \sqrt[n]{a^n} \notin R$ (n is an even number)

(*Example*):

$$\sqrt{0.25} + \sqrt{1.21} + \sqrt{1.44} = ?$$

(Solution):

$$\sqrt{0.5^2} + \sqrt{1.1^2} + \sqrt{1.2^2}$$

=0.5+1.1+1.2=2.8

(Example):

$$\frac{\sqrt[3]{(-3)^3} - \sqrt[6]{(-3)^6}}{-\sqrt{(-2)^2} - \sqrt[6]{-243}} = ?$$

A)-6   B)-3   C)3   D)6

(Solution):

$$\frac{\sqrt[3]{(-3)^3} - \sqrt[6]{(-3)^6}}{-\sqrt{(-2)^2} - \sqrt[6]{(-3)^5}}$$

$$= \frac{-3-3}{-2+3} = \frac{-6}{1} = -6$$

(Example):

$$\frac{\sqrt{2,7} + \sqrt{5,4}}{\sqrt{0,-1} + \sqrt{0,4}} = ?$$

(Solution):

$$\frac{\sqrt{2 + \dfrac{7}{9}} + \sqrt{5 + \dfrac{4}{9}}}{\sqrt{\dfrac{1}{9}} + \sqrt{\dfrac{4}{9}}}$$

$$= \frac{\sqrt{2+\dfrac{7}{9}} + \sqrt{5+\dfrac{4}{9}}}{\dfrac{1}{3}+\dfrac{2}{3}} = \frac{\dfrac{5}{3}+\dfrac{7}{3}}{\dfrac{3}{3}} = \frac{\dfrac{12}{3}}{3} = 4$$

PROPERTIES

1. $\sqrt[m]{a^n} = a^{\frac{n}{m}}$

(*Example*):

$$\sqrt[3]{3^{x+6}} + 9 \cdot \sqrt[3]{8.3^x} = 9 \quad \Rightarrow \quad x = ?$$

A)-3   B)-1   C)1   D)3   E)4

(Solution):

$$\sqrt[3]{3^x . 3^6} + 9 \cdot \sqrt[3]{8.3^x} = 9$$

$$3^{2} \sqrt[3]{3^x} + 2.9.\sqrt[3]{3^x} = 9$$

$$27\sqrt[3]{3^x} = 9$$

$$3^{\frac{x}{3}} = 3^{-1}$$

$$\frac{x}{3} = -1 \Rightarrow x = -3$$

-Answer A

2. $\sqrt[m]{\sqrt[n]{a}} = \sqrt[m.n]{a}$

$\sqrt[-x-3]{\sqrt[x-3]{81}} = \sqrt[4]{3} \Rightarrow x = ?$

A)1   B)3   C)5   D)7   E)9

(Solution):

$$\sqrt[x^2-9]{81} = \sqrt[4]{3}$$

$$3^{\frac{4}{x^2-9}} = 3^{\frac{1}{4}}$$

$$\frac{4}{x^2-9} = \frac{1}{4}$$

$$x^2 - 9 = 16 \Rightarrow x^2 = 25$$

X=5

4

$$3. \quad \sqrt{a + \sqrt{b}} = \sqrt{\frac{a + \sqrt{a^2 - b}}{2}} + \sqrt{\frac{a - \sqrt{a^2 - b}}{2}}$$

$$4. \quad \sqrt{a - \sqrt{b}} = \sqrt{\frac{a + \sqrt{a^2 - b}}{2}} - \sqrt{\frac{a - \sqrt{a^2 - b}}{2}}$$

(Example):

$$\sqrt{3 - 2\sqrt{2}} = ?$$

(Solution):

$$\sqrt{3 - 2\sqrt{2}} = \sqrt{(\sqrt{2} - \sqrt{1})^2}$$
$$= \sqrt{2} - 1$$

(Example):

$$(2 - \sqrt{3}) \cdot \sqrt{7 + 4\sqrt{3}} = ?$$

A)1   B)2   C)3   D)4   E)5

(Solution):

$$=(2-\sqrt{3}).\sqrt{7+2.2\sqrt{3}}$$
$$=(2-\sqrt{3}).\sqrt{7+2\sqrt{12}}$$
$$=(2-\sqrt{3}).(\sqrt{4}+\sqrt{3})$$
$$=(2)^2-(\sqrt{3})^2$$
$$=4\text{-}3=1$$

Answer A

(Example):

$$\sqrt{14+8\sqrt{3}}=2\sqrt{2}+b\Longrightarrow2\sqrt{3}.b=?$$

A) $\sqrt{2}$ B)$2\sqrt{2}$ C)$3\sqrt{2}$ D)$4\sqrt{2}$ E)$6\sqrt{2}$

(Solution):

$$\sqrt{14+2.4\sqrt{3}}=2\sqrt{2}+b$$
$$\sqrt{14+2\sqrt{48}}=2\sqrt{2}+b$$
$$\sqrt{8}+\sqrt{6}=\sqrt{8}+b$$
$$b=\sqrt{6}dir$$
$$2\sqrt{3}.b=2\sqrt{3}\sqrt{6}$$
$$=2\sqrt{18}=2.3\sqrt{2}=6\sqrt{2}$$

Answer E

(Example):

$$\sqrt{5}=x\Longrightarrow\sqrt{9-4\sqrt{5}}=?$$

6

A)x-1   B)x-2   C)x+1

D)x+2   E)x+4

(Solution):

$$\sqrt{9 - 2.2\sqrt{5}} = \sqrt{9 - 2\sqrt{20}}$$

$$= \sqrt{5} - \sqrt{4}$$

$$= x - 2$$

Answer B

(Example):

$$\sqrt{2 + \sqrt{3}} - \sqrt{4 - 2\sqrt{3}}$$

A)1   B)2   C)4   D)6   E)8

(Solution):

$$\sqrt{2 + \sqrt{3}} - \sqrt{4 - 2\sqrt{3}} = \sqrt{3} + 1 - (\sqrt{3} - 1)$$

$$= \sqrt{3} + 1 - \sqrt{3} + 1$$

$$= 2$$

(Example):

$$\sqrt{2 + \sqrt{3}} - \sqrt{2 - \sqrt{3}}$$

A) $\sqrt{2}$ B) $2\sqrt{2}$ C) 4   D) $4\sqrt{2}$ E) 8

(Solution):

$$\sqrt{2 + \sqrt{3}} - \sqrt{2 - \sqrt{3}}$$

$$= \left(\sqrt{\frac{3}{2}} + \sqrt{\frac{1}{2}} \right) - \left(\sqrt{\frac{3}{2}} - \sqrt{\frac{1}{2}} \right)$$

$$= \sqrt{\frac{3}{2}} + \sqrt{\frac{1}{2}} - \sqrt{\frac{3}{2}} + \sqrt{\frac{1}{2}}$$

$$= 2\sqrt{\frac{1}{2}} = \sqrt{4 \cdot \frac{1}{2}} = \sqrt{2}$$

-Answer A

5. $x = \sqrt[n]{a\sqrt[n]{a\sqrt[n]{a}}\ldots\ldots} \Rightarrow x = \sqrt[n-1]{a}$

6. $a.\sqrt[n]{x} + b.\sqrt[n]{x} - c.\sqrt[n]{x} = (a + b - c).\sqrt[n]{x}$

(Example):

$$\sqrt{40} + \sqrt{\frac{2}{5}} - \sqrt{\frac{2}{5}} = \frac{a\sqrt{40}}{20} \Rightarrow a = ?$$

A)9   B)11   C)13   D)15   E)17

(Solution):

$$\sqrt{40.\frac{2}{5}} + \sqrt{\frac{2}{5}.\frac{2}{5}} - \sqrt{\frac{5}{2}.\frac{2}{5}} = \frac{a\sqrt{40.\frac{2}{5}}}{20}$$

$$\sqrt{16} + \frac{2}{5} - \sqrt{1} = \frac{40.a}{20}$$

$$3 + \frac{2}{5} = \frac{a}{5}$$

$$a = 17$$

-Answer E

7. $\sqrt[n]{a}.\sqrt[n]{b} = \sqrt[n]{a.b}$

(Example):

$$\sqrt{2}\sqrt{3}.........\sqrt{n} = 2\sqrt{30} \Rightarrow n = ?$$

A)2   B)3   C)5   D)8   E)16

(Solution):

$$\sqrt{2}\sqrt{3}.........\sqrt{n} = \sqrt{4.30}$$

$$\sqrt{n!} = \sqrt{120}$$

$$l20 = 5l$$

$$n! = 5!$$

$$\Rightarrow n = 5$$

-Answer C

8. $\dfrac{\sqrt[n]{a}}{\sqrt[n]{b}} = \sqrt[n]{\dfrac{a}{b}}$

9. $\dfrac{a}{\sqrt{b}} = \dfrac{a.\sqrt{b}}{\sqrt{b}.\sqrt{b}} = \dfrac{a\sqrt{b}}{b}$

10. $\dfrac{a}{\sqrt{b}+\{\sqrt{c}} = \dfrac{a.(\sqrt{b}-\sqrt{c})}{(\sqrt{b}-\sqrt{c}).(\sqrt{b}+\sqrt{c})} = \dfrac{a.(\sqrt{b}-\sqrt{c})}{b-c}$

11. $\dfrac{a}{\sqrt{b}-\sqrt{c}} = \dfrac{a.(\sqrt{b}+\sqrt{c})}{(b-\sqrt{c}).(b+\sqrt{c})} = \dfrac{a.(b+\sqrt{c})}{b^2-c}$

(Example):

$$\frac{1}{\sqrt{3}-\sqrt{2}} - \frac{2}{\sqrt{2}} = ?$$

A) $\dfrac{\sqrt{3}}{2}$ B) $\sqrt{3}$ C) $2\sqrt{3}$ D) $3\sqrt{3}$ E) $4\sqrt{3}$

(Solution):

$$\frac{1}{\sqrt{3}-\sqrt{2}} - \frac{2}{\sqrt{2}}$$
$$= \frac{\sqrt{3}+\sqrt{2}}{3-2} - \frac{2\sqrt{2}}{2}$$
$$= \sqrt{3}+\sqrt{2}-\sqrt{2} = \sqrt{3}$$

-Answer B

(Example):

$$\frac{4}{\sqrt{5}-1} + \frac{1}{\sqrt{2}-1} - \frac{3}{\sqrt{5}-\sqrt{2}} = ?$$

A)1   B)2   C)3   D)4   E)5

(Solution):

$$\frac{4}{\sqrt{5}-1} + \frac{1}{\sqrt{2}-1} - \frac{3}{\sqrt{5}-\sqrt{2}}$$

$$= \frac{4(\sqrt{5}+1)}{5-1} + \frac{\sqrt{2}+1}{2-1} - \frac{3(\sqrt{5}+\sqrt{2})}{5-2}$$

$$= \sqrt{5}+1+\sqrt{2}+1-\sqrt{5}-\sqrt{2} = 2$$

-Answer B

TEST WITH SOLUTION

(Example):

1. $3\sqrt{147} + 2\sqrt{75} - 5\sqrt{108} = ?$

A)0   B) $\sqrt{3}$ C)-2$\sqrt{7}$ D) 2$\sqrt{7}$ E)6$\sqrt{3}$

Çözüm (Solution):

$$3\sqrt{147} + 2\sqrt{75} - 5\sqrt{108}$$
$$= 3\sqrt{49.3} + 2\sqrt{25.3} - 5\sqrt{36.3}$$
$$= 3.7\sqrt{3} + 2.5\sqrt{3} - 5.6\sqrt{3}$$
$$= 21\sqrt{3} + 10\sqrt{3} - 30\sqrt{3}$$
$$= \sqrt{3}$$

-Answer B

2. $3\sqrt{40} - \sqrt{250} + \frac{20}{\sqrt{10}} = ?$

A) $\sqrt{10}$ B) $3\sqrt{10}$ C) $5\sqrt{10}$ D) $2\sqrt{20}$ E) $\sqrt{50}$

(Solution):

$$3\sqrt{40} - \sqrt{250} + \frac{20}{\sqrt{10}}$$

$$= 3\sqrt{4.10} - \sqrt{25.10} + \frac{20.\sqrt{10}}{10}$$

$$= 3.2\sqrt{10} - 5.\sqrt{10} + 2.\sqrt{10}$$

$$= 6\sqrt{10} - 5\sqrt{10} + 2.\sqrt{10}$$

$$= (6 - 5 + 2)\sqrt{10} = 3\sqrt{10}$$

-Answer B

3. $18\sqrt{\dfrac{8}{27}} - \sqrt{150} = ?$

A)0   B) $-\sqrt{6}$ C)$\sqrt{6}$ D) 2-$\sqrt{3}$ E)2$\sqrt{3}$

(Solution):

$$18\sqrt{\frac{8}{27}} - \sqrt{150} = 18\sqrt{\frac{4.2}{9.3}} - \sqrt{25.3}$$

$$= 18.\frac{2\sqrt{2}}{3\sqrt{3}} - 5\sqrt{6}$$

$$= 12.\frac{\sqrt{2}}{\sqrt{3}} - 5\sqrt{6}$$

$$= 12.\frac{\sqrt{6}}{3} - 5\sqrt{6} = -\sqrt{6}$$

-Answer B

4. $\frac{1}{2}\sqrt{32} - \frac{1}{3}\sqrt{18} + \frac{\sqrt{6}}{\sqrt{3}} = ?$

A)0   B) -1 C)$\sqrt{2}$ D) $2\sqrt{2}$ E)$\sqrt{3}$

(Solution):

$$\frac{1}{2}\sqrt{32} - \frac{1}{3}\sqrt{18} + \frac{\sqrt{6}}{\sqrt{3}} = \frac{1}{2}\sqrt{16.2} - \frac{1}{3}\sqrt{9.2} + \frac{\sqrt{18}}{3}$$

$$= \frac{1}{2}.4\sqrt{2} - \frac{1}{3}.3\sqrt{2} + \frac{3\sqrt{2}}{3}$$

$$= 2\sqrt{2} - \sqrt{2} + \sqrt{2}$$

-Answer D

5. $\sqrt{\frac{1}{16} + \frac{1}{9}} . \sqrt{\frac{1}{9} - \frac{1}{25}} = ?$

A)1   B)$\frac{1}{3}$ C)16   D) $\frac{1}{9}$ E)$\frac{1}{20}$

$$\sqrt{\frac{1}{16}+\frac{1}{9}}.\sqrt{\frac{1}{9}-\frac{1}{25}} = \sqrt{\frac{25}{144}}.\sqrt{\frac{16}{225}}$$

$$= \frac{5}{12}.\frac{4}{15} = \frac{20}{180} = \frac{1}{9}$$

-Answer D

11. $4.\sqrt{\frac{3}{2}} - \sqrt{54} + 3\sqrt{\frac{2}{3}} = ?$

A)6   B)$\sqrt{6}$ C) $-\sqrt{6}$ D)0   E)$6\sqrt{6}$

(Solution):

$4.\sqrt{\frac{3}{2}} - \sqrt{54} + 3\sqrt{\frac{2}{3}}$

$= \frac{4.\sqrt{6}}{2} - \sqrt{9.6} + \frac{3.\sqrt{6}}{3}$

$= 2.\sqrt{6} - 3.\sqrt{6} + \sqrt{6} = 0$

-Answer D

12. $\sqrt{1.44} - \sqrt{19.6} + \sqrt{\frac{490}{25}} = ?$

$A)0.2$ B)0.9   C)1   D)1.2   E)1.4

(Solution):

$$\sqrt{\frac{144}{100}} - \sqrt{\frac{196}{10}} + \sqrt{\frac{490}{25}} = \frac{12}{10} \cdot \frac{14}{\sqrt{10}} + \frac{7\sqrt{10}}{5}$$

$$= \frac{12}{10} \cdot \frac{14 \cdot \sqrt{10}}{10} + \frac{7\sqrt{10}}{5}$$

$$= \frac{12}{10} - \frac{7\sqrt{10}}{5} + \frac{7\sqrt{10}}{5} = \frac{12}{10} = 1.2$$

-Answer D

13. $\left(1 - \frac{1}{\sqrt{2}}\right)\left(\frac{1}{2 - \sqrt{2}}\right) = ?$

(Solution):

$$\left(1 - \frac{1}{\sqrt{2}}\right) \cdot \left(\frac{1}{2 - \sqrt{2}}\right) = \left(1 - \frac{\sqrt{2}}{2}\right)\left(\frac{1}{2 - \sqrt{2}}\right)$$

$$= \frac{2 - \sqrt{2}}{2} \cdot \frac{1}{2 - \sqrt{2}}$$

$$= \frac{1}{2}$$

14. $$\frac{2}{1 - \dfrac{\sqrt{2}}{1 - \dfrac{1}{\sqrt{2}-1}}} = ?$$

A)-2   B)-1   C)0   D)1   E)2

Solution

$$\frac{2}{1 - \dfrac{\sqrt{2}}{1 - \dfrac{1}{\sqrt{2}-1}}} = \frac{2}{1 - \dfrac{\sqrt{2}}{1 - \dfrac{\sqrt{2}+1}{1}}}$$

$(\sqrt{2}+1)$

1- $\dfrac{\dfrac{\sqrt{2}}{1}}{1}$.

$$\frac{1}{1\text{-}\sqrt{2}-1} \frac{1}{1\text{-}\sqrt{2}-1}$$

$(\sqrt{2}+1)$

$$= \frac{2}{1 - \dfrac{\sqrt{2}}{1 - \sqrt{2} - 1}}$$

$$= \frac{2}{1 - \dfrac{\sqrt{2}}{-\sqrt{2}}}$$

$$= \frac{2}{1 + 1} = 1$$

-Answer

15. $\dfrac{\sqrt{45 + 20}}{4\sqrt{20} - \sqrt{5}}$

A) $\dfrac{5}{4}$ B) $\dfrac{5}{3}$ C) $\dfrac{5}{2}$ D) $\dfrac{5}{6}$ E) $\dfrac{5}{7}$

(Solution):

$$\frac{\sqrt{45} + \sqrt{20}}{4\sqrt{20} - \sqrt{5}} = \frac{\sqrt{9.5} + \sqrt{4.5}}{4\sqrt{4.5} - \sqrt{5}} =$$

$$\frac{3.\sqrt{5} + 2.\sqrt{5}}{8.\sqrt{5} - \sqrt{5}} = \frac{5.\sqrt{5}}{7\sqrt{5}} - = \frac{5}{7}$$

16. $\dfrac{\sqrt{3} + \sqrt{2}}{\sqrt{3} - \sqrt{2}} - 2\sqrt{6}$ =?

A)1   B)2   C)$\sqrt{6}$ D)4   E)5

18

(Solution):

$$\frac{\sqrt{3}+\sqrt{2}}{\sqrt{3}-\sqrt{2}} - 2\sqrt{6} = \frac{(\sqrt{3}+\sqrt{2})\,(\sqrt{3}+\sqrt{2}\,)}{(\sqrt{3}-\sqrt{2}\,)(\sqrt{3}-\sqrt{2}\,)} - 2\sqrt{6}$$

$$= \frac{3+\sqrt{66}+\sqrt{6}+2}{1} - 2\sqrt{6}$$

$$=5+2\sqrt{6}-2\sqrt{6}=5$$

-Answer E

17. $\sqrt{2}+\sqrt{3}.\dfrac{1}{\sqrt{2}+\sqrt{3}}$

A)-2$\sqrt{2}$ B)-$\sqrt{2}$ C) 2$\sqrt{2}$ D)3$\sqrt{2}$ E)4$\sqrt{2}$

(Solution):

$$\sqrt{2}+\sqrt{3}.\frac{1}{\sqrt{2}+\sqrt{3}}$$

$$(\sqrt{2}-\sqrt{3})$$

$$=\sqrt{2}+\sqrt{3}.\frac{\sqrt{2}-\sqrt{3})}{(\sqrt{2}+\sqrt{3})\,(\sqrt{2}-\sqrt{3})}$$

$$= \sqrt{2} + \sqrt{3} - \frac{(\sqrt{2} - \sqrt{3})}{-1}$$

$$= \sqrt{2} + \sqrt{3} + \sqrt{2} - \sqrt{3} = 2\sqrt{2}$$

-Answer C

18. $\dfrac{10}{\sqrt[3]{25}}$

A) $\sqrt{5}$ B) $\sqrt[3]{5}$ C) $2\sqrt[3]{5}$ D) -) $\sqrt[3]{5}$ E) $5\sqrt{5}$

(Solution):

$$\frac{10}{\sqrt[3]{25}} = \frac{10}{\sqrt[3]{5^2}} = \frac{10 \cdot \sqrt[3]{5}}{5} = 2\sqrt[3]{5}$$

$(\sqrt[3]{5})$ **-Answer C**

19. $\dfrac{1}{\sqrt{3}+1} - \dfrac{3}{\sqrt{3}-1} + \dfrac{3}{\sqrt{3}} = ?$

A)-$2\sqrt{3}$ B)-2   C)1   D)2
E) $2\sqrt{3}$

(Solution):

20

$$\frac{1}{\sqrt{3}+1} \cdot \frac{3}{\sqrt{3}-1} + \frac{3}{\sqrt{3}}$$

($\sqrt{3}$-1) $\sqrt{3}$+1) ($\sqrt{3}$)

$$= \frac{\sqrt{3}-1)}{3-1} - \frac{3.\sqrt{3}+1}{3-1} + \frac{3\sqrt{3}}{3}$$

$$= \frac{\sqrt{3}-1)}{2} - \frac{3.\sqrt{3}+1}{2} + \sqrt{3})$$

$$= \frac{\sqrt{3}-1-3\sqrt{3}-3}{2} + \sqrt{3}$$

$$= \frac{-2\sqrt{3}-4}{2} + \sqrt{3}$$

$$= \frac{2(-\sqrt{3}-2)}{2} + \sqrt{3}$$

$=-\sqrt{3}$**-2+**$\sqrt{3}$

=-2

-Answer B

20. $\dfrac{(\sqrt[3]{4}-\sqrt[3]{2})}{\sqrt[3]{2}-1}$ **=?**

A)2   B)4   C)6   D)8   E)10

(Solution):

$$\frac{(\sqrt[3]{4} - \sqrt[3]{2})}{\sqrt[3]{2} - 1} = \frac{\sqrt[3]{4^2} - \sqrt{2^2}}{\sqrt[3]{2} - 1}$$

$$= \frac{\sqrt[3]{16} - 2}{\sqrt[3]{2} - 1} = \frac{\sqrt[3]{2^3 \cdot 2} - 2}{\sqrt[3]{2} - 1}$$

$$= \frac{2\sqrt[3]{2} - 2}{\sqrt[3]{2} - 1}$$

$$= \frac{2(\sqrt[3]{2} - 1)}{\sqrt[3]{2} - 1} = 2 \qquad \text{-Answer B}$$

21. $\dfrac{\sqrt{\dfrac{0.4}{10}}}{\sqrt{0.04} - \sqrt{0.16}} = ?$

A)4   B)2   C)0.5   D)-1   E)-1.5

(Solution):

$$\frac{\sqrt{\dfrac{0.4}{10}}}{\sqrt{0.04}-\sqrt{0.16}}=\frac{\sqrt{0.04}}{\sqrt{0.04}-\sqrt{0.16}}$$

$$=\frac{0.2}{0.2-0.4}=\frac{0.2}{-0.2}\text{=-1}$$

-Answer D

22. $\dfrac{\sqrt{0.64}-\sqrt{1.96}}{\sqrt{0.36}}+1$ =?

A)-2　　　B)-1　　　C)0　　　D)1　　　E)2

(Solution):

$$\frac{\sqrt{0.64}-\sqrt{1.96}}{\sqrt{0.36}}+1=\frac{\sqrt{(0.8)^2}-\sqrt{(1.4)^2}}{\sqrt{(0.6)^2}}+1$$

$$=\frac{0.8-1.4}{0.6}+1$$

$$=\frac{-0.6}{0.6}+1\text{=-1+1=0}$$

-Answer C

23

23. $\sqrt{(0.6)^{-1} \cdot 6^{-1}} : (1.3)^{-1} =?$

A)3   B)2   C)$\dfrac{1}{2}$ D)$\dfrac{2}{3}$ E)$\dfrac{3}{8}$

(Solution):

$$\sqrt{(0.6)^{-1} \cdot 6^{-1}} \; : \; (1.3)^{-1} = \sqrt{\left(\frac{6}{9}\right)^{-1} \frac{1}{6}} \; : \; \left(\frac{13-1}{9}\right)^{-1}$$

$$=\sqrt{\frac{9}{6} \cdot \frac{1}{6} \cdot \frac{9}{12}}$$

$$\sqrt{\frac{9}{36}} \; : \; \frac{12}{9} = 6 : \frac{9}{12} = \frac{2}{3}$$

-Answer D

24. $\sqrt{(-2)^{-4}} + \left(\dfrac{1}{3}\right)^{-1} - \sqrt[3]{-8} =?$

A)5.25   B)-1.25   C)5   D)1.25   E)9

(Solution):

24

$$\sqrt{(-2)^{-4}} + \frac{(\frac{1}{3})^{-1}}{3} - \sqrt[3]{-8} = \sqrt{\frac{1}{(-2)^4}} + 3 - \sqrt[3]{-8}$$

$$\sqrt{\frac{1}{16}} + 3 - (-2)$$

$$= \frac{1}{4} + 3 + 2 = \frac{1}{4} + 5$$

$$= \frac{21}{4} = 5.25$$

-Answer D

25. $\dfrac{\sqrt{7 + \dfrac{1}{9}}}{0.13}$ =?

A)10   B)15   C)18   D)20   E)30

(Solution):

$$\frac{\sqrt{7 + \frac{1}{9}}}{0.13} = \frac{\sqrt{\frac{64}{9}}}{\frac{13-1}{90}} = \frac{\sqrt{\frac{8}{3}}}{\frac{12}{90}}$$

$$=\frac{8}{3}.\frac{90}{12} = 20$$

Yanit – Answer C

26 $\frac{2}{\sqrt[7]{8}}. = ?$ **A)**$\sqrt[7]{2}$ **B)** $\sqrt[7]{4}$ **C)** $\sqrt[7]{8}$ D)$\sqrt[7]{16}$
E) $\sqrt[7]{32}$

(Solution):

$$\frac{2}{\sqrt[7]{8}} = \frac{2}{\sqrt[7]{2^3}} = 2^{\frac{2}{3}}_{\frac{7}{7}} = 2^{1-\frac{3}{7}}$$

$$= 2^{\frac{4}{7}} = \sqrt[7]{2^4} = \sqrt[7]{16}$$

Yanit – Answer D

27. $\sqrt{2}.\sqrt[3]{3} = ?$

A)$\sqrt[6]{6}$ **B)** $\sqrt[6]{36}$ **C)** $\sqrt[6]{48}$ **D)** $\sqrt[6]{54}$

E) $\sqrt[6]{54}$

(Solution):

$$\sqrt{2}\cdot\sqrt[3]{3}=\sqrt[6]{2^3}\cdot\sqrt[6]{3^2}$$

$$=\sqrt[6]{8}\cdot\sqrt[6]{9}=\sqrt[6]{72}$$

-Answer E

28. $\dfrac{\sqrt{2}}{\sqrt{2}+\dfrac{1}{\sqrt{2}+\dfrac{1}{\sqrt{2}}}}=?$

A) $\dfrac{2}{3}$ B) $\dfrac{3}{4}$ C) $\dfrac{4}{5}$ D) $\dfrac{5}{4}$ E) $\dfrac{4}{3}$

(Solution):

$$\dfrac{\sqrt{2}}{\sqrt{2}+\dfrac{1}{\sqrt{2}+\dfrac{1}{\sqrt{2}}}}\quad \dfrac{\sqrt{2}}{\sqrt{2}+\dfrac{1}{\dfrac{3}{\sqrt{2}}}}\quad \dfrac{\sqrt{2}}{\sqrt{2}+\dfrac{\sqrt{2}}{3}}$$

$$=\dfrac{\sqrt{2}}{\dfrac{4\sqrt{3}}{3}}=\sqrt{2}\cdot\dfrac{3}{4\sqrt{2}}=\dfrac{3}{4}$$

-Answer B

29.
$$\frac{\sqrt{(-4)^2}+3\sqrt{9}-\sqrt{(-3)^2}}{\sqrt{(-1)^2}+\sqrt{16}}=?$$

A)1   B)2   C)3   D)4   E)5

(Solution):

$$\frac{\sqrt{(-4)^2}+3\sqrt{9}-\sqrt{(-3)^2}}{\sqrt{(-1)^2}+\sqrt{16}}=\frac{4+9-3}{1+4}$$

$$=\frac{10}{5}=2$$

-Answer B

$$\frac{\sqrt{20}+\sqrt{45}}{1\sqrt{8}+\sqrt{18}}=?$$

A) $\frac{3\sqrt{5}}{2}$ B)$2\sqrt{5}$ C) $\frac{\sqrt{10}}{2}$ D) $2\sqrt{10}$ E) $\frac{2\sqrt{10}}{3}$

(Solution):

$$\frac{\sqrt{20}+\sqrt{45}}{\sqrt{8}+\sqrt{18}} = \frac{2\sqrt{5}+3\sqrt{5}}{2\sqrt{2}+3\sqrt{2}} \frac{5\sqrt{5}}{=5\sqrt{2}} = \frac{\sqrt{5}}{\sqrt{2}} \frac{\sqrt{10}}{= 2}$$

– Answer C

$$2. \frac{2\sqrt{3}}{\sqrt{2}} + \frac{3\sqrt{2}}{\sqrt{3}} = ?$$

A)$2\sqrt{2}$ B) $2\sqrt{3}$ C) $3\sqrt{2}$ D) $2\sqrt{6}$ E) $2\sqrt{6}$

(Solution):

$$\frac{2\sqrt{3}}{\sqrt{2}} + \frac{3\sqrt{2}}{\sqrt{3}}$$

$$\frac{2\sqrt{3}}{\sqrt{2}} + \frac{3\sqrt{2}}{\sqrt{3}} = \frac{6}{\sqrt{6}} + \frac{6}{\sqrt{6}}$$

$$= \frac{12\sqrt{6}}{6}$$

$$= 2\sqrt{6}$$

3. $\sqrt{3^2} - \sqrt{(-3)^2} - (-2)(-3) = ?$

A)-6   B)0   C)3   D)6   E)12

(Solution):

$$\sqrt{3^2} = 3$$

$$\sqrt{(-3)^2} = |3| = 3$$

$$\sqrt{3^2} = \sqrt{(-3)^2}\text{-(-2).(-3)}$$

=3-3-(+6)=-6

-Answer A

4. $\dfrac{2^{1-n} \cdot \sqrt{8^n}}{\sqrt{2^{-n}}}$ =?

A)2^n B) 2^{n+1} C) 2^{-n}

D) $2^{\frac{1}{2}}$ E) 2^{-1}

(Solution):

30

$$\frac{2^{1-n}\cdot\sqrt{8^n}}{\sqrt{2^{-n}}} = \frac{2^{1-n}\cdot\sqrt{2^{3n}}}{2^{-\frac{n}{2}}}$$

$$\frac{2^{1-n}\cdot 2^{\frac{3n}{2}}}{2^{-\frac{n}{2}}}$$

$$2^{\left(1-n+\frac{3n}{2}+\frac{n}{2}\right)} = 2^{n+1}$$

-Answer B

5. $\sqrt{2^2}\cdot\sqrt{(-3)^2}\cdot\sqrt{(-3)^2}\cdot\sqrt{2^2}$ =?

A)-5   **B)-3**   **C)-1**   **D)1**   **E)2**

(Solution):

$$\sqrt{2^2}\cdot\sqrt{(-3)^2}\cdot\sqrt{(-3)^2}\cdot\sqrt{2^2}$$

$$\sqrt{2^2}=2$$

$$\sqrt{(-3)^2}=|-3|=3$$

$$=\sqrt{2^2}\cdot\sqrt{(-3)^2}\cdot\sqrt{(-3)^2}\cdot\sqrt{2^2}$$

=2.3-3-2

=6-5 =1

-Answer D

6.$\sqrt{(-8)^2} \cdot \sqrt[3]{(-8)^3}$ =?

A)-16   B)-8   C)0   D)8   E)18

(Solution):

$$\sqrt{(-8)^2} \cdot \sqrt[3]{(-8)^3}$$

$$\sqrt{(-8)^2} = |-8| = 8$$

$$\sqrt[3]{(-8)^3} = -8$$

$$\sqrt{(-8)^2} \cdot \sqrt[3]{(-8)^3} = 8-(-8)=16$$

-Answer

7.a=$\sqrt{5}$-1 $\Rightarrow \left(\dfrac{1}{a} - \dfrac{1}{b}\right)^{\frac{1}{2}}$ =?

   b=$\sqrt{5}$+1

A) $\dfrac{\sqrt{2}}{2}$ 　　　 B) $\dfrac{\sqrt{3}}{2}$ 　　 C)$2\sqrt{2}$ 　 D)$2\sqrt{3}$ 　　 E)$4\sqrt{2}$

(Solution):

$$\left(\frac{1}{a}-\frac{1}{b}\right)^{\frac{1}{2}}{}_{=}(\frac{1}{\sqrt{5}-1}-\frac{1}{\sqrt{5}+1})^{\frac{1}{2}}$$

$$\frac{1}{\sqrt{5}-1}=\frac{1}{\sqrt{5}-1.}\frac{\sqrt{5}+1}{\sqrt{5}+1}{}_{=}\frac{\sqrt{5}+1}{5-1}{}_{=}\frac{\sqrt{5}+1}{4}$$

$$_{=}\frac{1}{\sqrt{5}+1}=\frac{1}{\sqrt{5}+1.}\frac{\sqrt{5}-1}{\sqrt{5}-1}{}_{=}\frac{\sqrt{5}-1}{5-1}{}_{=}\frac{\sqrt{5}-1}{4}$$

$$_{=}(\frac{1}{\sqrt{5}-1}-\frac{1}{\sqrt{5}+1})^{\frac{1}{2}}{}_{=}(\frac{\sqrt{5}+1}{4}-\frac{\sqrt{5}-1}{4})^{\frac{1}{2}}$$

$$_{(}\frac{\sqrt{5}+1-\sqrt{5}+1}{4})^{\frac{1}{2}}{}_{=}(\frac{2}{4})^{\frac{1}{2}}{}_{=}\sqrt{\frac{1}{2}}{}_{=}\frac{1}{\sqrt{2}}=\frac{\sqrt{2}}{2}$$

　　　　　　　　　　　　　　　 -Answer B

8. $\dfrac{3}{\sqrt{7}-\sqrt{5}.}\dfrac{3}{\sqrt{7}+\sqrt{5}}=3p\Rightarrow p=?$

A)2 　　　 B)2 　　 C)$\sqrt{2}$ 　　 D) $\sqrt{3}$ 　 E) $\sqrt{5}$

(Solution):

$$\frac{3}{\sqrt{7}-\sqrt{5}}=\frac{3}{\sqrt{7}-\sqrt{5}}\cdot\frac{\sqrt{7}+\sqrt{5}}{\sqrt{7}+\sqrt{5}}$$

$$=\frac{3\sqrt{7}+3\sqrt{5}}{7-5}=\frac{3\sqrt{7}+3\sqrt{5}}{2}$$

$$\frac{3}{\sqrt{7}+\sqrt{5}}=\frac{3}{\sqrt{7}+\sqrt{5}}\cdot\frac{\sqrt{7}-\sqrt{5}}{\sqrt{7}-\sqrt{5}}$$

$$=\frac{3\sqrt{7}-3\sqrt{5}}{7-5}=\frac{3\sqrt{7}-3\sqrt{5}}{2}$$

$$\frac{3}{\sqrt{7}-\sqrt{5}}-\frac{3}{\sqrt{7}+\sqrt{5}}=\frac{3\sqrt{7}+3\sqrt{5}}{2}-\frac{3\sqrt{7}-3\sqrt{5}}{2}$$

$$\frac{3\sqrt{7}+3\sqrt{5}-3\sqrt{7}+3\sqrt{5}}{2}=$$

$$\frac{6\sqrt{5}}{2}=3\sqrt{5}=3p\Rightarrow p=\sqrt{5}$$

-Answer E

9. $\dfrac{\sqrt{0.81}+\sqrt{0.49}}{\sqrt{2.56}-\sqrt{1.44}}=?$

A)0.4   B)0.2   C)1   D)2   E)4

(Solution):

$$\frac{\sqrt{0.81} + \sqrt{0.49}}{\sqrt{2.56} - \sqrt{1.44}} = \frac{0.9 + 0.7}{1.6 - 1.2}$$

$$= \frac{1.6}{0.4} = 4$$

-Answer E

10. $\sqrt{3 + 2\sqrt{2}} - \sqrt{3 - 2\sqrt{2}} = ?$

A) $2\sqrt{3}$ **B)** $\sqrt{3}$ **C)** $\sqrt{2}$ **D)** 3 **E)** 2

(Solution):

$$\sqrt{3 + 2\sqrt{2}} - \sqrt{3 - 2\sqrt{2}} = \sqrt{(\sqrt{2} + 1)^2} - \sqrt{(\sqrt{2} - 1)^2}$$

$$= \sqrt{2} + 1 - (\sqrt{2} - 1)\sqrt{2} + 1 - \sqrt{2} + 1 = 2$$

-Answer E

11. a,b $\in Z$

$$\sqrt{72} - \sqrt{50} + \sqrt{27} = a\sqrt{2} + b\sqrt{3}$$

$$\Rightarrow 7a - b = ?$$

A)-3   B)-2   C)4   D)$2\sqrt{2}$ E)$3\sqrt{3}$

Çözüm **(Solution):**

$$\sqrt{72} - \sqrt{50} + \sqrt{27} = a\sqrt{2} + b\sqrt{3}$$

$$6\sqrt{2} - 5\sqrt{2} + 3\sqrt{3} = a\sqrt{2} + b\sqrt{3}$$

$$\sqrt{2} + 3\sqrt{3} = a\sqrt{2} + b\sqrt{3}$$

a=1$\Rightarrow$

b=3$\Rightarrow$ 7a-b=7-3=4

-Answer C

12.$\sqrt[3]{2} \cdot (3^{\sqrt[3]{32} - \sqrt[3]{108} + \dfrac{6}{\sqrt[3]{54}}})$ =?

A)$\sqrt[3]{4}$ B)$2\sqrt[3]{2}$ C)4   D)6   E)8

(Solution):

$$\sqrt[3]{2} \cdot (3^{\sqrt[3]{32} - \sqrt[3]{108} + \dfrac{6}{\sqrt[3]{54}}})$$

$$= \sqrt[3]{2} \cdot (3^{\dfrac{.2\sqrt[3]{4} - 3\sqrt[3]{4} + \dfrac{6}{3\sqrt[3]{2}}}})$$

$$= \sqrt[3]{2} \dfrac{\cdot(3.2\sqrt[3]{4} - 3\sqrt[3]{4} + 6)}{3\sqrt[3]{2}}$$

$$= \dfrac{18.2 - 9.2 + 6}{3}$$

$$= \dfrac{18 + 6}{3}$$

$$= \dfrac{24}{3} = 8 \qquad \textbf{-Answer E}$$

WORKBOOK TESTS

1. $5^{X+1} = \sqrt{25^{3X}} \Rightarrow X = ?$

A)0   B)1   C)$\dfrac{1}{2}$ D)$\dfrac{2}{3}$ E)4

2. $d^2 = \sqrt{2^{n+2}} \Rightarrow \sqrt[3]{d^6} = ?$

A)2^n B)$2^{\frac{1}{2}+n}$ C)$2^{\frac{n+2}{2}}$ D)2^{n-2} E)2^{3n+6}

3. $\sqrt{2010.1998 + 36} = ?$

A)1997   B)1999   C)2000

D)2002   E)2004

4. $\sqrt{x + \sqrt{x}} + \sqrt{x - \sqrt{x}} = 4 \Rightarrow x = ?$

A.$\dfrac{9}{4}$ B)$\dfrac{21}{8}$ C)$\dfrac{36}{11}$ D)$\dfrac{64}{15}$ E)$\dfrac{72}{18}$

$$5.\ x\cdot\sqrt{\dfrac{4}{3}}=\sqrt{\dfrac{3}{4}}+\sqrt{\dfrac{4}{3}}\Rightarrow x=?$$

A.$\dfrac{1}{2}$ B)$\dfrac{3}{5}$ C)$\dfrac{6}{7}$ D)$\dfrac{7}{4}$ E)$\dfrac{8}{5}$

$$6.\ \sqrt{3\cdot\sqrt[3]{3^{x}}}=\dfrac{1}{243}\Rightarrow x=?$$

A)-3   B)-14   C)-16   D)-18

E)-33

$$7.\ \sqrt{x+\sqrt{x^{2}}}\cdot\sqrt[3]{x+\sqrt{x^{2}}}=16^{\frac{5}{12}}\Rightarrow x=?$$

A)0   B)1   C)2   D)3   E)4

$$8.\ \dfrac{6-\sqrt{6}}{\sqrt{3}-\sqrt{2}}=?$$

A) $2\sqrt{3} - 2$ B) $4\sqrt{3} + 3\sqrt{2}$ C) $2\sqrt{2} - \sqrt{3}$ D))
$\sqrt{3} + \sqrt{2}$

E)) $\sqrt{2} - 3\sqrt{3}$

9. $\sqrt[X]{3^X\sqrt{729}} = 3 \Rightarrow X = ?$

A)0   B)1   C)2   D)3   E)4

10. X=$\sqrt[3]{\dfrac{2}{\sqrt[3]{2}}} \Rightarrow x^{18} = ?$

A)8   B)16   C)21   D)32   E)64

11. $\sqrt{\dfrac{3^{X+2}}{9^{X-1}}} = 27 \Rightarrow X = ?$

A)-5   B)-2   C)2   D)3   E)8

12. $$\frac{(\sqrt{5}-2).(\sqrt{9+2\sqrt{20}})}{\sqrt{2}}=?$$

A) $\frac{-\sqrt{2}}{3}$ B)1   C) $\frac{\sqrt{2}}{2}$ D) $\frac{-3}{2}$ E) $\frac{\sqrt{3}}{5}$

13. $\sqrt[4]{27\sqrt[4]{27\sqrt[4]{27}}} = X,$

$\sqrt{5\sqrt{5\sqrt{5}}} = Y \Rightarrow Y^2 - x^2 = ?$

A)8   B)12   C)16   D)21   E)27

14. $$\frac{1}{2-3\sqrt{3}} + \frac{1}{2+3\sqrt{3}} = ?$$

A) $-\frac{2}{3}$ B) $\frac{-14}{5}$ C) $\frac{-4}{23}$

D) $\frac{-8}{17}$ E) $\frac{\sqrt{3}}{2}$

15. $$\frac{5}{5-\sqrt{5}} \cdot (5+\sqrt{5})^{-1} = ?$$

A) $\dfrac{1}{2}$　　　B) $\dfrac{5}{2}$　　　C) $\dfrac{1}{4}$　　　D) $\dfrac{4}{7}$　　$\dfrac{7}{2}$

16. $\sqrt{9} + \sqrt{4} - \sqrt{(-4)^2} - \sqrt{(-2)^2} = ?$

A)1　　　　B)11　　　　　C)-10　　　　D)-11

E)-5

17. $\dfrac{1}{\sqrt{7-4\sqrt{3}}} + \dfrac{1}{2+\sqrt{3}} = ?$

A)2　　　B) $2\sqrt{3}$　　　　C)1　　　D)4　　　E) $\sqrt{3}$

18. $\sqrt{0.16} + \sqrt{0.64} = ?$

A) $\dfrac{3}{2}$　　　　　B) $\dfrac{6}{5}$　　　　　C) $\dfrac{4}{5}$

D) $\dfrac{9}{10}$　　　　　E) $\dfrac{12}{7}$

19. $\sqrt{\dfrac{3^{-1}}{0.3} : \dfrac{0.09}{10}}$ =?

A) $\dfrac{10}{3}$

B) $\dfrac{10}{9}$

C) $(\dfrac{9}{10})^{-2}$

D) $(\dfrac{3}{10})^{-2}$

E) $\dfrac{3}{5}$

20. $\sqrt{6 + \sqrt{6 + \sqrt{6 + \sqrt{6 + \dots\dots}}}} \sqrt{X + \sqrt{X + \sqrt{X + \dots\dots}}} \Rightarrow$

$\Rightarrow n = ?$

A)8   B)12   C)16   D)27   E)64

21. $\dfrac{1}{\sqrt{2}-1} - \dfrac{1}{1+\sqrt{2}} = ?$

A)0   B)$\sqrt{2}$ C)1   D)-1   E)2

22. $3^X = 8 \Rightarrow (9^X)^{-1} \cdot (81^X)^2 = ?$

A) a^{-2} B) a^2 C) a^4 D) a^{-5}

E) a^5

23. $\sqrt{a} + \dfrac{1}{\sqrt{a}}\sqrt{4} \Rightarrow a^2 + \dfrac{1}{a^2} = ?$

A)9   B)8   C)4   D)2   E)1

24. $\sqrt{3^4\sqrt{3^{2X}}} = (\dfrac{1}{81})^2 \Rightarrow$ X=?

A)-27   B)81   C)30   D)36   E)-21

(Answers)						
1.C	2.C	3.E	4.D	5.D	6.E	
7.C	8.B	9.D	10.B	11.B	12.C	
13.C	14.C	15.C	16.A	17.D	18.B	
19.D	20.D	21.E	22.E	23.D	24.C	

$1. 4\sqrt{8} + 5\sqrt{18} - 3\sqrt{72} + \sqrt{50} = ?$

A) $6\sqrt{2}$ B) $7\sqrt{2}$ C) $8\sqrt{2}$ D) $9\sqrt{2}$

E) $10\sqrt{2}$

$2. \sqrt{108} - \sqrt{48} - \sqrt{75} = ?$

A) $\sqrt{3}$ B) $-3\sqrt{3}$ C) $2\sqrt{3}$ D) $-\sqrt{3}$
E) 0

$3. 3\sqrt[3]{2} + 4\sqrt[3]{16} - 4\sqrt[3]{54} = ?$

A) $-2\sqrt[3]{2}$ B) $-\sqrt[3]{3}$ C) $-\sqrt[3]{2}$ D) $\sqrt[3]{3}$

E) $2\sqrt[3]{3}$

$4. \sqrt[3]{0.006} + \sqrt[3]{0.002} = ?$

A) $\dfrac{\sqrt[3]{2}(\sqrt[3]{2} + 1)}{10}$ B) $\dfrac{\sqrt[3]{2}(\sqrt[3]{3} + 1)}{10}$ C) $\dfrac{\sqrt[3]{3}(\sqrt[3]{2} - 1)}{10}$

D) $\dfrac{\sqrt[3]{3}(\sqrt[3]{2}+1)}{10}$ E) $\sqrt[3]{5}+\sqrt{2}$

5. $\sqrt{1-\dfrac{9}{25}}+\sqrt{1-\dfrac{11}{36}}=?$

A) $\dfrac{37}{25}$ B) $\dfrac{27}{25}$ C) $\dfrac{49}{30}$

D) $\dfrac{51}{25}$

E) $\dfrac{49}{16}$

6. $\sqrt{4^2-3^2}\cdot\sqrt[4]{7^2}=?$

A)0   B)1   C)2   D)3   E)40

7. $\dfrac{\sqrt[3]{16}+\sqrt[3]{54}-\sqrt[3]{250}+\sqrt[3]{128}}{\sqrt[3]{16}-\sqrt[3]{250}}=?$

A) $-\dfrac{1}{2}$ B) $-\dfrac{2}{3}$ C) $-\dfrac{3}{4}$

D) $-\dfrac{4}{3}$ E) $-\dfrac{5}{4}$

8.$3\sqrt{2} + 4\sqrt{8} - 5\sqrt{50} + 8\sqrt{32} = ?$

A)$10\sqrt{2}$ B) $12\sqrt{2}$ C) $4\sqrt{3}$

D) $16\sqrt{2}$ E) $18\sqrt{2}$

9.$2\sqrt[3]{3} - 3\sqrt[3]{24} + 4\sqrt[3]{81} = ?$

A)$\sqrt[3]{3}$ B)$2\sqrt[3]{3}$ C)$4\sqrt[3]{3}$

D)$6\sqrt[3]{3}$ E) $8\sqrt[3]{3}$

10.$X > 0, Y > 0, Z > 0 \Rightarrow$

$6\sqrt{XY^2Z^2} + \dfrac{8}{2}\sqrt{XY^2Z^4} - \dfrac{6}{Y}\sqrt{XY^4Z^2} = ?$

A)$8YZ\sqrt{X}$ B) $6YZ\sqrt{X}$ C) $\sqrt{X^2 - 1}$

47

D) $\sqrt{X-1}$ E)$6Y\sqrt{X-1}$

11.$X > 1 \Rightarrow \sqrt{(X+1)^3} - X\sqrt{X+1} + \sqrt{(X-1)(X^2-1)} = ?$

A) $X\sqrt{X+1}$ B) $\sqrt{X+1}$ C)$\sqrt{X^2-1}$

D) $\sqrt{X-1}$ E) $X\sqrt{X-1}$

12.$\sqrt{\dfrac{3}{2}} + \sqrt{\dfrac{2}{3}} = ?$

A) $\dfrac{\sqrt{3}}{6}$ B)$6\sqrt{\dfrac{6}{5}}$ C)$3\dfrac{\sqrt{6}}{2}$

D)$\dfrac{5}{\sqrt{6}}$ E)$2\sqrt{6}$

13.$\sqrt{0.04} - 2\sqrt[3]{0.008} - \sqrt[4]{0.0016} + \sqrt{1.69} = ?$

A)0.5   B)0.6   C)0.7   D)0.8

E)0.9

14. $\dfrac{1}{\sqrt[3]{a^2}} = ?$

A)$\sqrt[3]{a^2}$ B)$\sqrt[3]{a}$ C) $\dfrac{\sqrt[3]{a}}{a}$

D) $\dfrac{\sqrt[3]{a^2}}{a}$ E)$\sqrt{a}$

15. $\sqrt{a}.\sqrt[3]{b}.\dfrac{1}{\sqrt[3]{ab}} = ?$

A)1   B) $\sqrt{a}$ C) $\sqrt[3]{ab}$

D) $\sqrt[6]{a}$ E) $\sqrt[6]{ab}$

16. $\sqrt{2\sqrt[3]{2\sqrt{2}}} = ?$

A) $\sqrt[12]{2^8}$ B) $\sqrt[6]{2^5}$ C) $\sqrt[12]{2^5}$

D) $\sqrt[4]{2^3}$ E) $\sqrt[16]{2^8}$

17. $\sqrt[4]{\dfrac{x^3}{\sqrt[3]{x^2}}} : x^{\frac{7}{12}} = ?$

A)0   B)1   C)2   D)3   E)x^4

18. $\dfrac{\sqrt[3]{5}.\sqrt{2}}{\sqrt[6]{10}} = ?$

A)$\sqrt[6]{10}$ B) $\sqrt[5]{15}$ C) $\sqrt[5]{20}$

D) $\sqrt[6]{25}$ E) $\sqrt[5]{200}$

19. $\sqrt[2]{2}.\sqrt[3]{2}.\sqrt{2} = ?$

A)$2\sqrt[3]{5}$ B) $2^{30}\sqrt{2}$ C) $2^{15}\sqrt{3}$

D) $2^{30}\sqrt{7}$ E) $2^{30}\sqrt{3}$

20. $\sqrt[4]{2\sqrt[3]{2\sqrt{2}}} = ?$

A) $\sqrt[16]{8}$ B) $\sqrt[4]{4}$ C) $\sqrt[8]{8}$

D) $\sqrt[24]{8}$ E) $\sqrt[12]{6}$

21. $\dfrac{\sqrt{300} - 2\sqrt{27}}{\sqrt{75} + \sqrt{3}} = ?$

A)1   B)$\dfrac{1}{2}$ C) $\dfrac{2}{3}$

D)$2\sqrt{3}$ E)$3\sqrt{3}$

(Answers)						
1.E	2.B	3.C	4.B	5.C	6.A	
7.D	8.E	9.E	10.A	11.A	12.D	
13.E	14.C	15.D	16.D	17.B	18.C	
19.B	20.C	21.B				

1. $\sqrt{\sqrt{0.0036} + \sqrt{0.09}} \cdot \dfrac{10}{\sqrt{2}} = ?$

A)3   B)$5\sqrt{2}$ $\sqrt{3}$

D) $3\sqrt{2}$ E)5

2. $\dfrac{\sqrt[6]{(64)^{-1}}}{\sqrt[3]{4} \cdot \sqrt[4]{4}} \cdot \sqrt[6]{2} = ?$

A)$\dfrac{1}{2}$ B)2   C)8   D) $\dfrac{1}{4}$

E) $\sqrt[6]{2}$

3. $\dfrac{\sqrt{32} - \sqrt{45} + \sqrt{2} - \sqrt{20}}{\sqrt{5} - \sqrt{2}} = ?$

A)-5   B)$\sqrt{5} + \sqrt{2}$ C)4

D)$\sqrt{2}$ E)$3\sqrt{5}$

4. $\dfrac{a}{\sqrt[n]{a^{n-1}}} = ?$

A)$\sqrt[n]{a}$ B)$\sqrt[n]{a^1}$ C)$\sqrt[n]{a^{n+1}}$

D)$\sqrt[1]{a}$ E)$\sqrt[n-1]{a}$

5. $\sqrt{48} - \sqrt{12} - \sqrt{\dfrac{4}{3}}.\sqrt{\dfrac{1}{3}} = ?$

A)0   B)1   C)$\sqrt{3}$

D)$\dfrac{1}{3}\sqrt{3}$ E)$\dfrac{2}{3}\sqrt{3}$

6.$\sqrt{1.21} - $**35.**$\sqrt[3]{0.008} - $**10.**$\sqrt[4]{\dfrac{0.0016}{(0.25)^2}} = ?$

A)0   B)$\dfrac{1}{2}$ C)$\dfrac{2}{3}$ D)5   E)11

7.2-$\left[3:(\sqrt{3:2})^2 - 2:(\sqrt{2:3})^2\right]$-**3**=?

A)-3   B)-2   C)0

D)$\dfrac{2}{3}$ E)$\dfrac{3}{2}$

8. $\left(\sqrt[9]{\sqrt[3]{27x^6}}\right)^2 \cdot \left(\sqrt[6]{\sqrt[4]{x^8}}\right)^2 = ?$

A) X　　　**B)** $2\sqrt{X}$　　　**C)** $3\sqrt{X}$

D) $X\sqrt{X}$　　　**E)** $2x^2$

9. $\sqrt{18} - \sqrt[3]{16} - 3\sqrt{2} + \sqrt[3]{54} = a^3\sqrt{2} = ?$

A) -1　　　**B)** 0　　　**C)** 1

D) $\sqrt[3]{2}$　　　**E)** $\sqrt{2}$

10. $\dfrac{5\sqrt{\dfrac{1}{2}} - -\sqrt{0.5} + \sqrt{200}}{\sqrt{8}} = ?$

A)12　　　**B)** $6\sqrt{2}$　　　**C)6**

D)4　　　**E)** $2\sqrt{2}$

11 . $4\sqrt{45} - 2\sqrt{80} - \sqrt[4]{25} = ?$

$A)3\sqrt{5}$　　　**B)** $2\sqrt{5}$　　**C)** $\sqrt{5}$

$D)0$　　　　　**E)** $9\sqrt{5}$

12. $\sqrt{45} - 10\sqrt{\dfrac{1}{5}} + \sqrt{80} - \sqrt[4]{25}$ =?

A) $3\sqrt{5}$ **B)** $7\sqrt{5}$ **C)** $4\sqrt{5}$

D) $5\sqrt{5}$ **E)** $2\sqrt{5}$

13. $3\sqrt{\dfrac{4}{3}} - 2\sqrt{\dfrac{25}{3}} + \sqrt{\dfrac{49}{3}} = ?$

A) $\sqrt{3}$ **B)** $2\sqrt{3}$ **C)** $3\sqrt{3}$

D) $\dfrac{\sqrt{3}}{3}$ **E)** $5\sqrt{3}$

14. $\dfrac{\sqrt{5.76} + \sqrt{2.89} + \sqrt{1.96}}{\sqrt{0.49} + \sqrt{0.15}} = ?$

A) $\dfrac{1}{2}$ **B)** 2   **C)** 3

D) 4   **E)** 5

15. $\sqrt{8.1} + \sqrt{4.9} - \sqrt{12.1} = ?$

A)$\sqrt{10}$ B)$2\sqrt{10}$ C) $\dfrac{\sqrt{10}}{2}$

D) $\dfrac{\sqrt{10}}{5}$ E)5

16. $\sqrt{\dfrac{4}{25} - \dfrac{3}{5} + \dfrac{9}{16}} = ?$

A)-1   B)2   C) $\dfrac{1}{3}$

D) $\dfrac{7}{20}$ E) $\dfrac{1}{4}$

17. $\dfrac{7}{\sqrt{7} - \dfrac{3}{\sqrt{7} - \dfrac{3}{\sqrt{7}}}} = ?$

A)1   B) $\dfrac{1}{4}$ C)$4\sqrt{7}$

D)4   E) $2\sqrt{7}$

18. $3\sqrt{0.64} + 8\sqrt{0.49} = ?$

A)4   B)5   C)6   D)7   E)8

19. $\sqrt[3]{\dfrac{6}{7^{1-3X}} + \dfrac{7^{3X}}{7}} = ?$

A)7^{2X} B) 7^{3X} C) 7^X

D)7   E)49

20. $\dfrac{\dfrac{\sqrt{252}}{\sqrt{7}}}{\sqrt{7} + \sqrt{\dfrac{1}{3}}} \cdot \dfrac{\sqrt{27}}{} = ?$

A)1   B)3   C)6

D)9   E)15

21 $\sqrt{0.21 + \sqrt{0.0016}} + \sqrt{0.53 - \sqrt{0.000064}} = ?$

A)1   B)0.3   C)1.1

D)1.2   E)1.5

22.$4\sqrt{2.52} - 2\sqrt{3.43} = ?$

A)$\sqrt{7}$ **B)**$2\sqrt{7}$ **C)** $3\sqrt{7}$

D)0   **E)**1

23.$\sqrt[5]{0.008} : \sqrt[5]{25} = ?$

A)1   **B)**0.1   **C)**0.2   **D.**0.6   **E)**$\dfrac{1}{50}$

24.$\sqrt[3]{0.5}.\sqrt[6]{0.25}.\sqrt[12]{0.0625} = ?$

A)1   **B)** $\dfrac{1}{2}$ **C)** $\dfrac{1}{4}$ **D)**5   **E)** $\dfrac{1}{5}$

(Answers)					
1.D	2.D	3.A	4.B	5.C	6.A
7.C	8.E	9.C	10.C	11.A	12.C
13.A	14.E	15.C	16.D	17.C	18.E
19.C	20.E	21.D	22.A	23.C	24.B

1. $(4\sqrt{5} - 2\sqrt{3})^2 - (4\sqrt{5} + 2\sqrt{3})^2 = ?$

A) $32\sqrt{15}$　　　B) $16\sqrt{15}$　　　C) $-20\sqrt{15}$

D) $-24\sqrt{15}$　　　E) $-32\sqrt{15}$

2. $\sqrt{6 - \sqrt{6 - \sqrt{6 - \sqrt{6......}}}} = ?$

A)1　　B)2　　C)3　　D)6　　E)36

3. $\sqrt{3\sqrt{3\sqrt{3.......}}} = a \Rightarrow a^2 = ?$

A)3　　B)6　　C)9　　D)12　　E)36

4. $\sqrt[3]{49 : \sqrt[3]{49 : \sqrt[3]{49.........}}} X$

$\sqrt{4\sqrt{4\sqrt{4..........}}} = Y \Rightarrow x^2 - Y = ?$

A)3　　B)4　　C)7　　D)45　　E)53

59

5. $\sqrt{6 + \sqrt{6 + \sqrt{6\ldots\ldots}}} = ?$

A-2　　B)-3　　C)0

D)2　　E)3

6. $\sqrt{7 - 2\sqrt{12}} + \sqrt{8 + 2\sqrt{15}} = ?$

A)$2+\sqrt{5}$　　B) $2-\sqrt{5}$　　C) $\sqrt{5}+1$　　D) $2\sqrt{3}$

E) $\sqrt{3}$

7. $\sqrt{5 + 2\sqrt{6}} + \sqrt{8 - 2\sqrt{15}} - \sqrt{9 - 4\sqrt{5}} = ?$

A) $3+\sqrt{3}$　　B) $5\sqrt{2}$　　C) $2+\sqrt{2}$

D) $4+\sqrt{5}$　　E) $2\sqrt{3}$

8. $\dfrac{4}{\sqrt{10 + 2\sqrt{21}}} + \sqrt{3} = ?$

A) $4\sqrt{3}$　　B) $5\sqrt{2}$　　C) $6\sqrt{7}$

D) $\sqrt{7}$　　E) $2\sqrt{3}$

9. $\sqrt{5 + 2\sqrt{4}} \cdot \sqrt{5 - 2\sqrt{4}} = ?$

A)2   B)3   C)4

D)9   E)16

10. $\dfrac{1}{\sqrt{6 - 2\sqrt{8}}} - \dfrac{1}{\sqrt{6 + 2\sqrt{8}}} = ?$

A)$\sqrt{2}$ B) $\sqrt{3}$ C)2 $\sqrt{3}$

D) $2\sqrt{2}$ E) $3\sqrt{2}$

11. $\sqrt{\dfrac{\sqrt{3} + 1}{\sqrt{3} - 1}} - \sqrt{\dfrac{\sqrt{3} - 1}{\sqrt{3} + 1}} = ?$

A. $2\sqrt{3}$ B) $-\sqrt{3}$ C) $-2\sqrt{3}$

D) $\sqrt{2}$ E)3

12. $\dfrac{\sqrt{3} + \sqrt{5}}{\sqrt{3} - \sqrt{5}} - \dfrac{\sqrt{3} - \sqrt{5}}{\sqrt{3} + \sqrt{5}} = ?$

A) $-\sqrt{15}$ B) $-2\sqrt{15}$ C) $-3\sqrt{15}$

D) $-4\sqrt{15}$ E) $\sqrt{15}$

13. $\sqrt{2\sqrt{\dfrac{1}{X}}} = 2 \Rightarrow X = ?$

A) 2^{-6} B) 2^{-5} C) 2^{-4}

D) 2^{4} E) 2^{5}

14. $\dfrac{\sqrt[8]{16}\cdot\sqrt[4]{0.25}}{\sqrt{0.1}} = ?$

A) 10   B) $2\sqrt{5}$ C) $\sqrt{6}$

D) $\sqrt{10}$ E) $4\sqrt{5}$

15. $\sqrt[a]{\dfrac{9^{a+1} - 3^{2a}}{8,3^{a}}}$ =?

A) 1   B) 2   C) 3

D)3^a **E)** 3^{-a}

16. $\sqrt[8]{2\sqrt{2}} = a \Rightarrow \sqrt[3]{a^{16}} = ?$

A)1   **B)**2   **C)**4

D)8   **E)**16

17. $\dfrac{1}{\sqrt{0.08}} + \dfrac{2}{\sqrt{0.32}} = ?$

A)1   **B)**$\sqrt{2}$ **C)**10

D)$5\sqrt{2}$ **E)**$2\sqrt{3}$

18. $\sqrt{3\sqrt[3]{X}} = \sqrt[3]{2\sqrt{3}} \Rightarrow X = ?$

A)3   **B)**6   **C)**27

D)$\dfrac{4}{9}$ **E)**$\dfrac{3}{2}$

19. $\dfrac{\sqrt{6X} - \sqrt{3X}}{\sqrt{2} - 2} = -\sqrt{6} \Rightarrow X?$

A)$\sqrt{2}$ **B)**$\sqrt{5}$ **C)**4

D)2   **E)**$\sqrt{22}$

20. $(\sqrt{3}+\sqrt{2})^{100} \cdot (5-\sqrt{24})^{50} = ?$

A)-1   **B)**1   **C)**3^{40}

D) 2^{100} **E)** $(\sqrt{3}+\sqrt{2})^{50}$

21. $\sqrt[3]{5\sqrt[4]{5}\sqrt[3]{5}} = 5^X \Rightarrow X = ?$

A)$\dfrac{4}{9}$ **B)** $\dfrac{4}{5}$ **C)** $\dfrac{8}{9}$

D) $\dfrac{6}{5}$ **E)**2

22. $\sqrt{8-27-\sqrt{48}} = ?$

A) $\sqrt{3} - 1$ **B)** $\sqrt{3} + 1$ **C)** $\sqrt{3} + \sqrt{2}$

D) $\sqrt{3} + 2$ **E)** $2\sqrt{3} + 2$

(Answers)					
1.A	2.D	3.A	4.B	5.B	6.E
7.E	8.C	9.C	10.D	11.E	12.A
13.C	14.D	15.C	16.B	17.D	18.D
19.C	20.E	21.A	22.B		

13. $\dfrac{\sqrt{36}}{\sqrt[4]{81}}$=?

A)5   B)4   C)3   D)2   E)1

(Solution)

$$\dfrac{\sqrt{36}}{\sqrt[4]{81}} = \dfrac{\sqrt{6^2}}{\sqrt{3^4}} = \dfrac{6}{3} = 2$$

-Answer D

14. $\dfrac{\sqrt[3]{128} + \sqrt[3]{16}}{\sqrt[3]{2}} = ?$

A)$\sqrt[3]{2}$ B)$2\sqrt[3]{2}$ C)$3\sqrt[3]{2}$ D)3   E)6

(Solution)

$$\dfrac{\sqrt[3]{128} + \sqrt[3]{16}}{\sqrt[3]{2}} = \dfrac{\sqrt[3]{2^6}\cdot\sqrt[3]{2} + \sqrt[3]{2^3}\cdot\sqrt[3]{2}}{\sqrt[3]{2}}$$

$$\dfrac{4\sqrt[3]{2} + 2\sqrt[3]{2}}{\sqrt[3]{2}} = 6\dfrac{\sqrt[3]{2}}{\sqrt[3]{2}} = 6$$

-Answer E

15. $x^a=\sqrt{5} \Rightarrow x^{-4a} = ?$

A) $\dfrac{1}{125}$ B) $\dfrac{1}{25}$ C) $\dfrac{1}{5}$ D)5   E)25

Cozum(Solution)

$x^a=\sqrt{5} \Rightarrow x^{-4a}=(x^a)^{-4}$

$(\sqrt{5})^{-4}$

$\dfrac{1}{(\sqrt{5})^{-4}} = \dfrac{1}{25}$

-Answer B

16. $\dfrac{\sqrt{10}}{\sqrt{2}+\sqrt{6}} + \dfrac{\sqrt{10}}{\sqrt{6}-\sqrt{2}} = ?$

A)$\sqrt{2}$ B) $\sqrt{3}$ C) $\sqrt{5}$ D) $\sqrt{15}$ E)$2\sqrt{15}$

(Solution)

$\dfrac{\sqrt{10}}{\sqrt{2}+\sqrt{6}} + \dfrac{\sqrt{10}}{\sqrt{6}-\sqrt{2}}$

$(\sqrt{5} - \sqrt{2})$ $(\sqrt{5} + \sqrt{2})$

$$= \frac{\sqrt{60} + 2\sqrt{10} + \sqrt{60} - 2\sqrt{10}}{6 - 2}$$

$$= \frac{2\sqrt{60}}{4}$$

$$= \frac{8\sqrt{15}}{4} = 2\sqrt{15}$$

-Answer E

17. $\sqrt{-X + 2\sqrt{X - 1}} + \sqrt{Y - \sqrt{2Y - 1}} = 0 \Rightarrow X + Y = ?$

A)2   B)3   C)4   D)5   E)6

(Solution)

$$\sqrt{-X + 2\sqrt{X - 1}} + \sqrt{Y - \sqrt{2Y - 1}} = 0$$

$\Rightarrow -X + 2\sqrt{X - 1} = 0$ $Y - \sqrt{2Y - 1} = 0$

$2\sqrt{X - 1} = 0$ $Y = \sqrt{2Y - 1}$

$X = 2\sqrt{X - 1}$ $(Y)^2 = (\sqrt{2Y - 1})^2$

$(X)^2 = (2\sqrt{X - 1})^2$ $Y^2 = 2Y - 1$

$X^2 = 4X - 4$ $Y^2 - 2Y + 1 = 0$

$X^2 - 4X + 4 = 0$ $(Y - 1)^2 = 0$

$\Rightarrow X = 2 \Rightarrow X + Y = 2 + 1 = 3$ $\Rightarrow Y = 1$

1. $\dfrac{\sqrt{aa}.\sqrt{bb}}{\sqrt{a}\sqrt{b}} = ?$

A)$11\sqrt{a}$ B)$11b$ C)10

D)$\sqrt{a}$ E)11

2. a,b $\in$ R

a.b=16$\Rightarrow \sqrt[4]{a\sqrt{b}}.\sqrt[4]{b\sqrt{a}} = ?$

A)$\sqrt{2}$ B)$2\sqrt{2}$ C)$2 + \sqrt{3}$

D) $4 + \sqrt{3}$ E)8

3. $\dfrac{\sqrt{3} - 2}{1 + \sqrt{2}} = P \Rightarrow \dfrac{1 - \sqrt{2}}{2 + \sqrt{3}} = ?$

A)p   B)$\sqrt{3}P$ C)2P

D)3P   E)4P

4.$\sqrt{2\sqrt[3]{X}} = 2\sqrt{2} \Rightarrow X = ?$

A)4   B)8   C)16

D)32   E)64

5) $\dfrac{X - \sqrt{45} + 20}{\sqrt{180 - X}} = 4 \Rightarrow X = ?$

A)$8\sqrt{5}$ B) $5\sqrt{5}$ C) $2\sqrt{8}$

D) $\sqrt{5}$ E)2

6.$\dfrac{\sqrt[3]{(-4)^3}}{\sqrt{(-4)^2}} + \dfrac{\sqrt{(-7)^2}}{\sqrt{49}} = ?$

A)0   B)1   C)2   D)3   E)6

7.$3^X + 3^{X-2} = 30 \Rightarrow \sqrt[x]{0.125} = ?$

70

A) $\dfrac{1}{5}$ B) $\dfrac{\sqrt[3]{3}}{5}$ C) $\dfrac{1}{2}$

D)) $\dfrac{1}{8}$ E)1

8. $\sqrt[4]{3\sqrt[3]{X}} = \sqrt[4]{27\sqrt[3]{3}} \Rightarrow X = ?$

A) $3^{\frac{1}{12}}$ B) $3^{\frac{1}{3}}$ C) 3^3

D) 3^8 E)3

9. $5^a = x$

$\sqrt{x^2\sqrt{x}} = 25 \Rightarrow a = ?$

A) $\dfrac{1}{5}$ B)1   C) $\dfrac{8}{5}$

D)2   E)8

10. $(5\sqrt{2} + 2\sqrt{3})^2 = X + Y\sqrt{150} \Rightarrow X + Y = ?$

A)16   B)32   C)48   D)54

E)66

11. $\sqrt{36X + 36} + \sqrt{9X + 9} = 18 \Rightarrow X = ?$

A)2   B)3   C)4

D)5   E)6

12. $\sqrt[n]{16^6 + 8^8} \in z \Rightarrow \min(n) = ?$

A)16   B)20   C)24

D)25   E)10

13. $\sqrt[4]{\dfrac{1}{81} + \dfrac{1}{144} - \dfrac{1}{54}}?$

A) $\dfrac{2}{3}$ B) $\dfrac{1}{3}$ C) $\dfrac{1}{6}$

D) $\dfrac{1}{9}$ E) $\dfrac{1}{2}$

14. $\sqrt{2^{X+3} + 2^X} = 48 \Rightarrow X = ?$

A)4   B)5   C)6

D)7   E)8

15. $X + a = \sqrt{a^2 + 6} \Rightarrow$ x.y=?

$\text{y-a} = \sqrt{a^2 + 6}$

A)6   B)9   C)12

D)15   E)24

16. $\sqrt{4^{X-1}} \cdot \sqrt{2^X} = 16 \Rightarrow X = ?$

A) $\sqrt[3]{2}$ B)2   C)4

D) $\sqrt{2}$ E) $2\sqrt{2}$

17. $\sqrt[3]{a\sqrt{a}} = 2 \Rightarrow \sqrt{a\sqrt{a}} = ?$

A) $\sqrt{2}$ B)3   C) $2\sqrt{2}$

D)4   E) $3\sqrt{2}$

18. $\sqrt{22.5} + \sqrt{8.1} = a\sqrt{10} \Rightarrow a = ?$

A) $\dfrac{12}{5}$
B) $\dfrac{11}{5}$
C) $\dfrac{3}{5}$

D)2
E)3

19. $(\sqrt{2} - 1)^2.(\sqrt{6} + \sqrt{3})^2 = ?$

A)3
B)2
C)1

D)$\sqrt{2}$
E)$\sqrt{3}$

20. $\dfrac{a}{b} - \dfrac{b}{a} = 4 \Rightarrow \sqrt{\dfrac{a^4 + b^4}{a^2 b^2}} + 46 = ?$

A)6
B)7
C)8

D)9
E)10

21. $\sqrt{\dfrac{15}{4^{1-a}}} + 4^{a-1} = 64 \Rightarrow a = ?$

A)2   B)3   C)4   D)5   E)6

22. $X+Y=0 \Rightarrow (X + \sqrt{X^2 + 1}) \cdot (Y + \sqrt{Y^2 + 1}) = ?$

A)1   B) $x^2 - 1$ C) $2x^2 - 1$

D)-1   E) $2x^2$

23. $X, Y \in R$

$\sqrt{X - 2Y} + \sqrt{Y + 2} = 0 \Rightarrow Z = ?$

X+Y-Z=8

A)-16   B)-14   C)-12

D)-6   E)-3

(Answers)					
1.E	2.B	3.A	4.E	5.B	6.A
7.C	8.E	9.C	10.E	11.B	12.D
13.C	14.E	15.A	16.C	17.A	18.A
19.A	20.C	21.D	22.A	23.	

1. $\dfrac{\sqrt{6}+\sqrt{2}}{\sqrt{6}-\sqrt{3}+\sqrt{2}-1} \cdot \dfrac{2}{\sqrt{2}} = ?$

A)2   B)$\sqrt{2}$ C) $1+\sqrt{2}$

D)$2+\sqrt{2}$ E)4

2. $\sqrt{(1+\sqrt{5})^2} \cdot \sqrt{6-2\sqrt{5}} = ?$

A)$-\sqrt{5}$ B)-1   C)1

D)14   E)9

3. $X,Y \in R$

$\sqrt{\dfrac{X}{Y}} + \sqrt{\dfrac{Y}{X}} = 3X + 3Y \Rightarrow X.Y = ?$

A)$\dfrac{1}{9}$ B) $\dfrac{1}{3}$ C)1

D)3   E)9

$4. \sqrt[3]{24 + \sqrt[3]{X\sqrt[3]{8}}} = 3 \Rightarrow X = ?$

A)21   B)25   C)27

D)34   E)40

$5. \dfrac{\sqrt[3]{4^{X+1}}}{\sqrt[3]{8^{X-1}}} = 16 \Rightarrow X = ?$

A) $-\dfrac{13}{5}$ B) $-\dfrac{11}{5}$ C) $-\dfrac{9}{5}$

D) $\dfrac{11}{5}$ E) $\dfrac{13}{5}$

$6. \sqrt[x]{0.16} = a \Rightarrow a^{x+1} = ?$

A) $\dfrac{1}{6}$ B)3   C) $\dfrac{5a}{3}$

D) $\dfrac{a}{3}$ E) $\dfrac{a}{6}$

$7. \dfrac{1}{\sqrt{2}} + \dfrac{\sqrt{5 + 2\sqrt{3 + \sqrt{9}}}}{\sqrt{6} + 2} = ?$

A)$3\sqrt{2}$ B)3   C) $2\sqrt{2}$

D)2   E)$\sqrt{2}$

8.$X=\dfrac{1}{\sqrt{3}+\sqrt{2}} \Rightarrow 5-2\sqrt{6}=?$

A)10X   B)10^{X^2} C)X^2

D) 2^{X^2} E) 3^{X^2}

9.$a=1-\sqrt{3} \Rightarrow \sqrt{a^2}-\sqrt{(b-a)^2}+\sqrt[3]{-a^3}=?$

   $b=\sqrt{2}-1$

A)-2   B)-1   C)$\sqrt{3}-\sqrt{2}$

D)$2-\sqrt{3}$ E)$\sqrt{2}+\sqrt{3}$

10. $\dfrac{\sqrt{144}}{0.6}+\dfrac{\sqrt{2.56}}{0.2}-\dfrac{\sqrt{0.64}}{0.4}=?$

A)2   B)4   C)6

D)8   E)10

11. $\sqrt[3]{a\sqrt{b}} = \sqrt[6]{432} \Rightarrow a = b = ?$

A)5   B)7   C)9

D)12   E)15

12. $\sqrt{2} = 1.41 \Rightarrow \sqrt{18} + \sqrt{27} = ?$

$\sqrt{3} = 1.73$

A)9.42   B)9.48   C)10.12

D)10.32   E)10.56

www.ingramcontent.com/pod-product-compliance
Lightning Source LLC
Chambersburg PA
CBHW072203170526
45158CB00004BB/1751